Keep Your Mind Fit

Numbers

CARLTON BOOKS

WHAT IS MENSA?

Mensa is the international society for people with a high IQ.
We have more than 100,000 members in over 40 countries worldwide.

The society's aims are:
> to identify and foster human intelligence for the benefit of humanity
> to encourage research in the nature, characteristics, and uses of intelligence
> to provide a stimulating intellectual and social environment for its members

Anyone with an IQ score in the top two per cent of population is eligible to become a member of Mensa – are you the 'one in 50' we've been looking for?

Mensa membership offers an excellent range of benefits:
> Networking and social activities nationally and around the world
> Special Interest Groups – hundreds of chances to pursue your hobbies and interests – from art to zoology!
> Monthly members' magazine and regional newsletters
> Local meetings – from games challenges to food and drink
> National and international weekend gatherings and conferences
> Intellectually stimulating lectures and seminars
> Access to the worldwide SIGHT network for travellers and hosts

For more information about Mensa: www.mensa.org, or

British Mensa Ltd.,
St John's House,
St John's Square,
Wolverhampton
WV2 4AH

Telephone: +44 (0)1902 772771
E-mail: enquiries@mensa.org.uk
www.mensa.org.uk

Contents

INTRODUCTION 6

PUZZLES 8

ANSWERS 156

INTRODUCTION

Puzzles are as old as humankind. It's inevitable - it's the way we think. Our brains make sense of the world around us by looking at the pieces that combine to make up our environment. Each piece is then compared to everything else we have encountered. We compare it by shape, size, colour, textures, a thousand different qualities, and place it into the mental categories it seems to belong to. Then we consider other nearby objects, and examine what we know about them, to give context. We keep on following this web of connections until we have enough understanding of the object of our attention to allow us to proceed in the current situation. We may never have seen a larch before, but we can still identify it as a tree. Most of the time, just basic recognition is good enough, but every time we perceive an object, it is cross-referenced, analysed, pinned down - puzzled out.

This capacity for logical analysis - for reason - is one of the greatest tools in our mental arsenal, on a par with creativity and lateral induction. Without it, science would be non-existent, and mathematics no more than a shorthand for counting items. In fact, although we might have

INTRODUCTION

made it out of the caves, we wouldn't have got far.

Furthermore, we automatically compare ourselves to each other – we place ourselves in mental boxes along with everything else. We like to know where we stand. It gives us an instinctive urge to compete, both against our previous bests and against each other. Experience, flexibility and strength are acquired through pushing personal boundaries, and that's as true of the mind as it is of the body. Deduction is something that we derive satisfaction and worth from, part of the complex blend of factors that goes into making up our self-image. We get a very pleasurable sense of achievement from succeeding at something, particularly if we suspected it might be too hard for us.

Number puzzles are always popular and offer a particular type of challenge: we are dealing with a finite universe and nothing is left to chance.

So enjoy the puzzles in this book – and why not try one of the others in the Keep Your Mind Fit series?

Happy Puzzling!

PUZZLE 1

Insert the missing numbers. In each pattern the missing number has something to do with the surrounding numbers in combination.

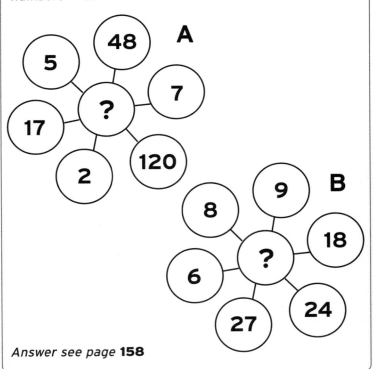

Answer see page **158**

PUZZLE 2

If Picasso is worth 28
and Monet is worth 22,
how much is
Raphael worth?

Answer see page **158**

PUZZLE 3

Take a five-digit number and reverse it. Subtract the original number from its reverse, and you are left with 33957. What was the original number?

Answer see page **158**

PUZZLE 4

What number should replace the question mark?

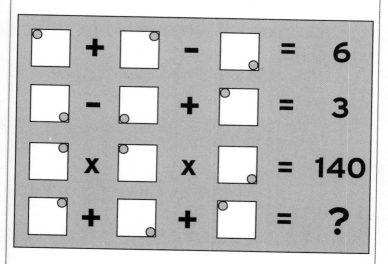

Answer see page **158**

PUZZLE 5

Za-za is older than Fifi, but younger than Juan. Fifi is older than Jorjio and Maccio. Maccio is younger than both Carlos and Jorjio. Juan is older than both Fifi and Maccio, but younger than Carlos.
Who is the oldest, and who is the youngest?

Answer see page 158

PUZZLE 6

A rectangular swimming pool of constant depth is twice as long as it is wide, but the owner is unhappy with the dimensions of the pool. The length is reduced by 12 units and its width increased by 10 units. When this is done, the modified pool will hold exactly the same volume of water. What were the pool's original dimensions?

Answer see page **158**

PUZZLE 7

Each shape is made up of two items, and each same shape has the same value, whether in the foreground or background. What number should replace the question mark?

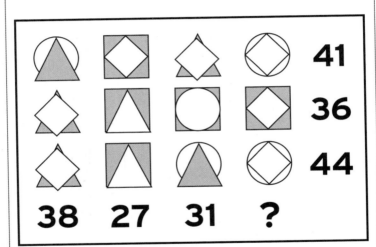

Answer see page **158**

PUZZLE 8

What is the area of the shaded path, if the path is one unit wide?

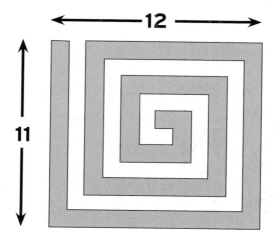

Answer see page **102**

PUZZLE 9

The panel below, when complete, contains the binary numbers from 1 to 25. Does binary patch A, B, C or D complete the panel?

1	1	0	1	1	1	0	0	1	0	1
1	1	0	1	1	1	1	0	0	0	1
0	0	1	1	0	1	0	1	0	1	1
1	1	0						1	1	1
0	1	1			?			0	0	1
0	0	0						0	1	0
0	1	1	1	0	1	0	0	1	0	1
0	1	1	0	1	1	0	1	0	1	1
1	1	1	0	0	0	1	1	0	0	1

A

1	0	1	1	1
1	1	1	1	0
1	1	1	0	0

B

0	1	1	0	1
1	1	1	0	0
0	1	0	0	1

C

1	1	0	1	1
1	1	0	1	1
0	0	1	0	1

D

0	1	1	0	1
1	1	1	0	0
1	1	0	0	1

Answer see page **158**

PUZZLE 10

Which letters, based on the alphanumeric system, should go into the blank boxes?

6	1	7	3			
1	3	5	4	A	H	B
7	7	0	9			

5	1	3	9			
2	8	6	4	F	B	C
8	6	2	6			

2	2	9	2			
4	3	0	9			
7	1	7	8			

Answer see page **158**

PUZZLE 11

What number, when you multiply it by 5 and add 6, then multiply that result by 4 and add 9, gives you a number that, when you multiply it by 5 and subtract 165, gives you a number that, when you knock off the last 2 digits, brings you back to your original number?

Answer see page **159**

PUZZLE 12

What number should replace the question mark?

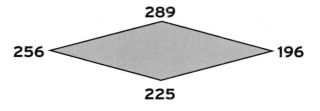

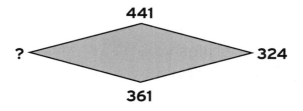

Answer see page **159**

PUZZLE 13

If each large ball weighs one and a third times the weight of each little ball, what is the minimum adjustment that will make the scales balance?

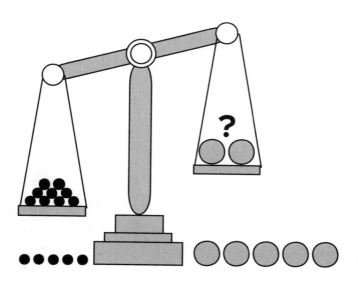

Answer see page **159**

PUZZLE 14

Present at Juan's birthday party were a father-in-law, a mother-in-law, a daughter-in-law, two sons, two daughters, two sisters and a brother, four children, three grandchildren, two fathers, two mothers, a grandfather, and a grandmother. However, family relationships can be complicated. One man's brother can, of course, be another man's brother-in-law, and at the same time, someone's son. With that in mind, what is the smallest number of people needed at the party for the above relationships to exist?

Answer see page **159**

PUZZLE 15

How many rosettes are missing from the blank circle?

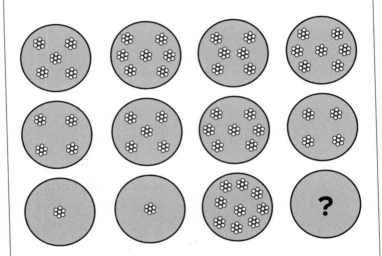

Answer see page **159**

PUZZLE 16

Forty people took part in a freestyle race. Twenty people ran. Ten people dashed. Five people bolted and sprinted. Three people bolted, dashed, ran and sprinted. Two people ran, bolted, and sprinted. Five people ran and sprinted. Two people dashed, ran, and sprinted. How many people neither dashed, ran, bolted, nor sprinted?

Answer see page **159**

PUZZLE 17

What value needs to go into the upper box to bring this system into balance? Note: The beam is broken down into equal parts and the value of each box is taken from its midpoint.

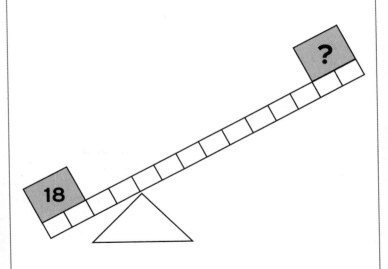

Answer see page **159**

PUZZLE 18

Using only the numbers already used, complete this puzzle to make all the rows, columns, and long diagonals add to 27.

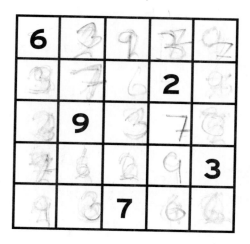

Answer see page **159**

PUZZLE 19

Insert the supplied rows of numbers into the appropriate places in the grid to make all rows, columns and long diagonals add to 175.
Example: (C) goes into the location (a).

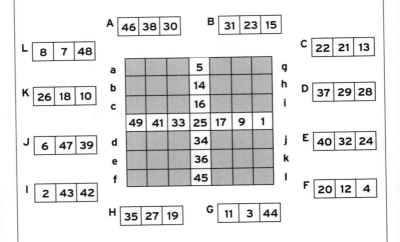

Answer see page **159**

PUZZLE 20

At 3pm one day, a flagpole and a measuring pole cast shadows as shown. What length is the flagpole?

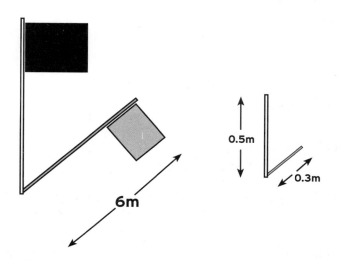

Answer see page **159**

PUZZLE 21

Use logic to discover which shape has the greatest perimeter.

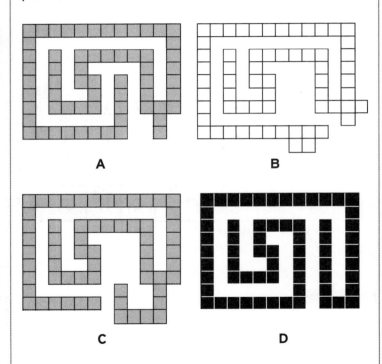

A

B

C

D

Answer see page **160**

PUZZLE 22

Crack the code to find the missing number.

A	B	C	D	E	F	G	H	I	J
9	3	8	7	8	9	2	8	5	7
1	2	1	5	?	7	1	0	1	2
K	L	M	N	O	P	Q	R	S	T

Answer see page **160**

PUZZLE 23

What number should replace the question mark?

6 8 4 8
7 9 6 ?

Answer see page **160**

PUZZLE 24

What number replaces the question mark?
What is the value of each animal?

19   15   18   22

Answer see page 160

PUZZLE 25

What number should replace the question mark?

A	B	C	D	E
3	11	7	4	18
2	12	7	5	19
5	17	11	6	?

Answer see page **160**

PUZZLE 26

If it takes 5 men to dig 5 holes in 5 hours, how many men does it take to dig 100 holes in 100 hours?

Answer see page **160**

PUZZLE 27

Put the right number in the blank star.

Answer see page **160**

PUZZLE 28

If you buy 9 barrels of beer for 25 Credits each, but you are given a 25% discount on the last 4 barrels, what was the total cost of the barrels?

Answer see page **160**

PUZZLE 29

When a ball is dropped from a height of 9m, it bounces back two-thirds of the way. Assuming that the ball comes to rest after making a bounce which takes it less than 2mm high, how many times does it bounce?

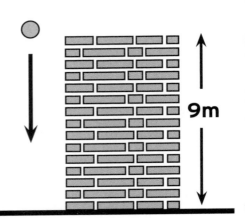

Answer see page **160**

PUZZLE 30

The planet Pento is inhabited by a race of highly intelligent one-toed quadrupeds with elephant-like trunks. So with four toes and a trunk, they have adopted the five base for their number system. With that in mind, convert the Pento number 1234 into its decimal equivalent.

Answer see page **160**

PUZZLE 31

What number should replace the question mark?

❄ + 🪔 − ☀ = 6

❄ × 🪔 × ☀ = 30

❄ − 🪔 − ☀ = 0

❄ + 🪔 + ☀ = ?

Answer see page **160**

PUZZLE 32

Each same shape has the same value. What number should replace the question mark?

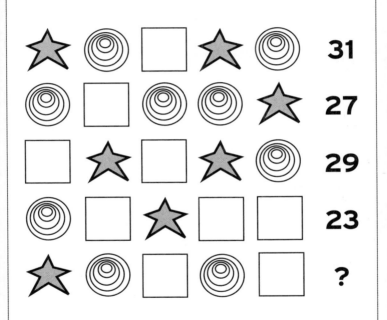

Answer see page **160**

PUZZLE 33

Find the missing number.

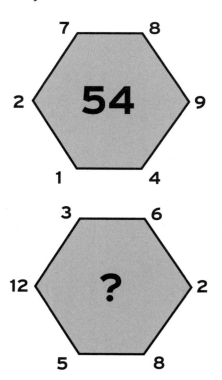

Answer see page **160**

PUZZLE 34

What three-digit number should replace the question mark?

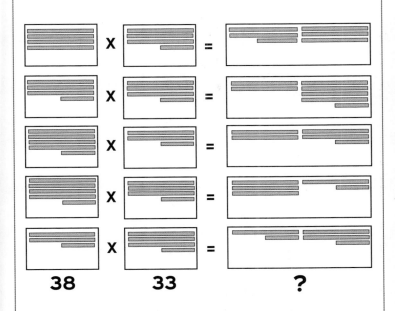

38   33   ?

Answer see page **161**

PUZZLE 35

The three balls at the top of each hexagon should contain numbers that, when added together and subtracted from the total of the numbers in the three balls at the bottom of each hexagon, equal the number inside each relevant hexagon. Insert the missing numbers.

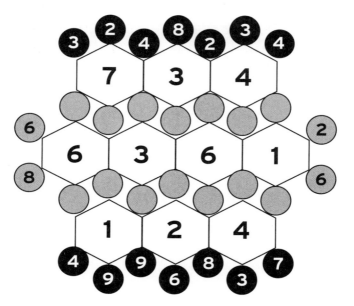

Answer see page **161**

PUZZLE 36

What number, when added to a number 10 times as big, gives a number that, when its right-hand digit is multiplied by four and added to the result of the above, gives 1000?

Answer see page 161

PUZZLE 37

What number should replace the question mark?

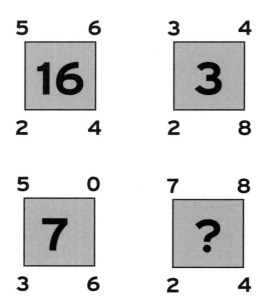

Answer see page **161**

PUZZLE 38

This clock has been designed for a planet that rotates on its axis once every 16 hours. There are 64 minutes to every hour, and 64 seconds to the minute. At the moment, the time on the clock reads a quarter to eight. What time, to the nearest second, will the clock say the time after the next time the hands appear to meet?

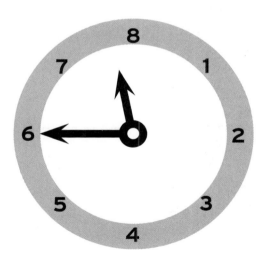

Answer see page **161**

PUZZLE 39

A large sheet of paper is 0.1mm thick. A man amuses himself by tearing it in half and putting both pieces together, and then tearing those into four sheets, and repeating the process until he has done it twenty-five times. How high is the stack of paper now?

a) As thick as a book

b) As high as a man

c) As high as a house

d) As high as a mountain

*Answer see page **161***

PUZZLE 40

This is a time puzzle. Which symbol is missing?
Is it A, B, C, D, E or F?

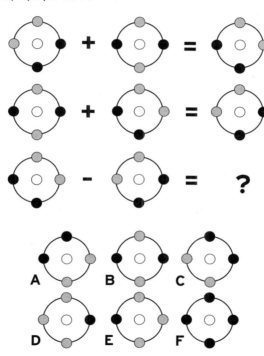

Answer see page **161**

PUZZLE 41

What number should replace the question mark?

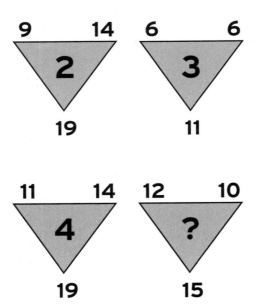

Answer see page **161**

PUZZLE 42

Insert in the boxes at the corner of each shaded number-square, the digits which are multiplied together to give the numbers in the shaded boxes. For example, in the bottom left corner, 144 is derived from 3 x 6 x 8 (and another multiplier – here 1), but you also have to consider how this helps to make solutions for the surrounding numbers... and so on.

3		5		4		4		3		3
	90		120		64		144		54	
2										1
	48		96		16		72		36	
1										2
	160		80		20		150		30	
4										1
	180		10		40		100		15	
9										3
	27		8		32		12		81	
3										9
	24		28		84		45		135	
8										1
	144		42		63		225		25	
3		6		1		3		5		1

Answer see page **162**

PUZZLE 43

What number should replace the question mark?

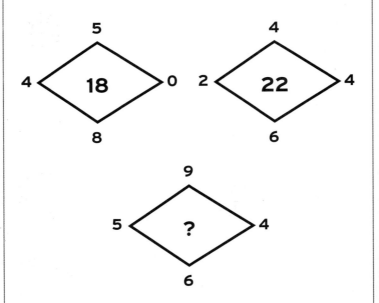

Answer see page **162**

PUZZLE 44

Each like symbol has the same value. Supply the missing total.

29   **37**   **24**   **?**

Answer see page **162**

PUZZLE 45

What time will it be, to the nearest second, when the hands of this clock next appear to meet?

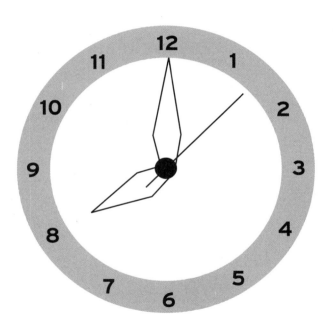

Answer see page **162**

PUZZLE 46

What number should replace the question mark?

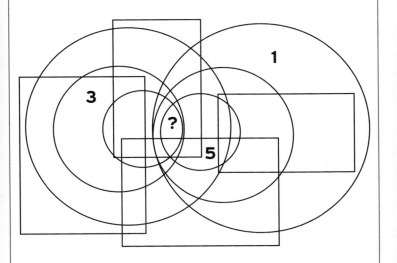

Answer see page **162**

PUZZLE 47

Insert the missing numbers in the blank hexagons.

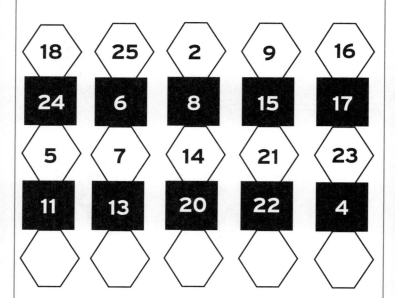

Answer see page **162**

PUZZLE 48

What number should replace the question mark?

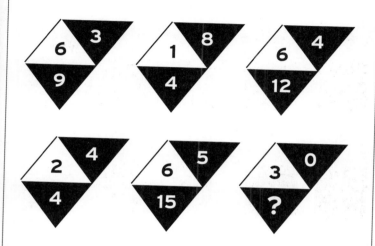

Answer see page **162**

PUZZLE 49

What number should replace the question mark?

9	7	2	5	7	4	3	2	5	1
									4
9	4	5	2	7	5	2	7		5
3							9		9
6		?	2	6	5	1	8		8
2									1
8	3	5	2	7	4	3	3	6	5

Answer see page **162**

PUZZLE 50

Black counters are nominally worth 4.
White counters are nominally worth 3.
Being on a diagonal trebles a counter's value.
Being on or in the innermost box doubles a counter's value.
Being on the outermost box halves a counter's value.
The rules work in combination.
What is the total value of all the counters on the board?

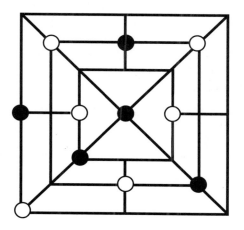

Answer see page **162**

PUZZLE 51

What number continues the sequence?

15	20	20
6	6	?

Answer see page **163**

PUZZLE 52

I have a deck of cards from which some are missing. If I deal them equally between nine people, I have two cards to spare. If I deal them equally between four people, I have three cards to spare. If I deal them between seven people, I have five cards to spare. There are normally 52 cards in the deck.

How many are missing?

Answer see page **163**

PUZZLE 53

Each same symbol has the same value. What number should replace the question mark?

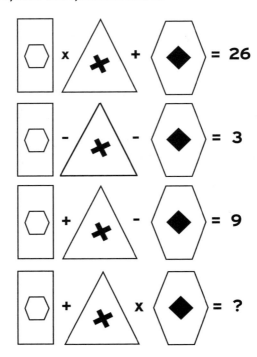

Answer see page **163**

PUZZLE 54

What number should replace the question mark?

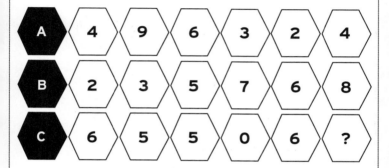

Answer see page **163**

PUZZLE 55

What number should replace the question mark in the blank square?

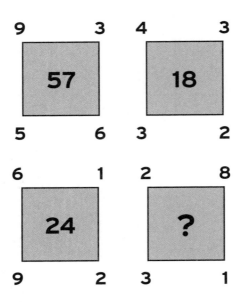

Answer see page **163**

PUZZLE 56

Insert the central numbers.

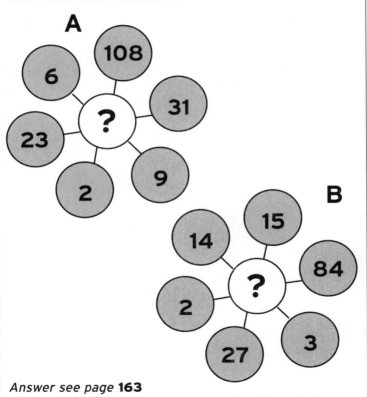

Answer see page 163

PUZZLE 57

What number should replace the question mark?

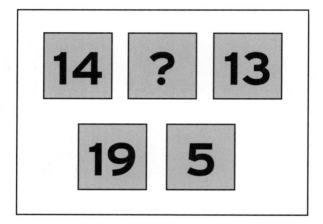

Answer see page **163**

PUZZLE 58

The symbols represent the numbers 1 to 9.
Work out the value of the missing multiplier.

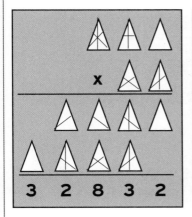
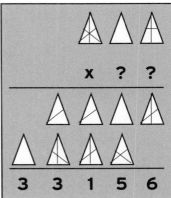

Answer see page **163**

PUZZLE 59

This system is balanced. How heavy is the black box (ignoring leverage effects)?

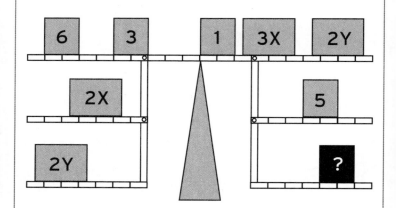

Answer see page **163**

PUZZLE 60

$$2 \times \sqrt{2} = \sqrt{8}$$

$$3 \times \sqrt{5} = \sqrt{45}$$

What number should replace the question mark?

$$4 \times \sqrt{6} = \sqrt{?}$$

Answer see page 163

PUZZLE 61

Somewhere within the numbers below left, there is a number that, if it is put into the grid opposite, starting at the top left and working from left to right, row by row, will have the middle column as shown when the grid is completed. Put in the missing numbers.

3095867235697809123948566809

4164162223456341219183621614

4432708929846152955001621932

0002528131215858719394504639

5123161762113267792289656123

1022384046128985404326161425

2616093417285830091242859648

1342568309980128473061339021

Answer see page **164**

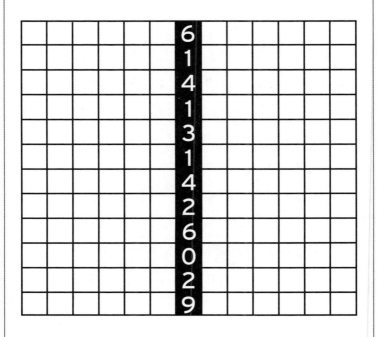

PUZZLE 62

How many different ways is it possible to arrange the order of these four kings?

Answer see page **164**

PUZZLE 63

If the top left intersection is worth 1, and the bottom right intersection is worth 25, which of these nodule grids, A, B, C or D, is worth 67?

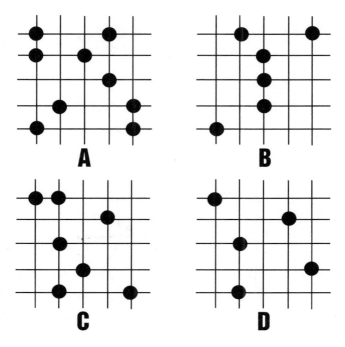

Answer see page 164

PUZZLE 64

Prior to the time shown, when were all four of the digits on this watch last on display?

Answer see page **164**

PUZZLE 65

Find within the number below, two numbers, one of which is double the other, and which when added together make 10743.

57162383581

Answer see page **164**

PUZZLE 66

This system is balanced. How heavy is the black weight (ignoring leverage effects)?

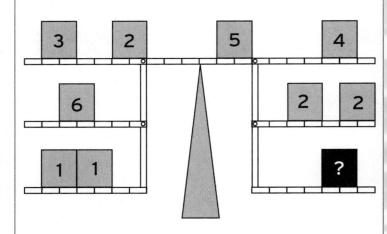

Answer see page **164**

PUZZLE 67

There are logical differences in the way each of these squares work, but they all involve simple addition or subtraction of rows. What are the missing numbers?

2	6	3	0	8	a
3	8	0	3	9	b
2	3	?	5	7	c
1	9	2	5	4	d
2	1	5	3	6	e

A

2	1	3	2	0	a
1	3	5	6	2	b
0	5	?	4	7	c
2	9	6	3	0	d
1	0	2	9	9	e

B

3	1	2	0	9	a
6	1	4	6	2	b
2	8	?	1	9	c
4	9	6	5	7	d
7	1	3	3	3	e

C

3	3	6	4	7	a
3	3	6	1	1	b
1	1	?	2	0	c
3	4	1	0	6	d
2	1	9	3	2	e

D

Answer see page **164**

PUZZLE 68

What number should replace the question mark in the third hexagon pattern?

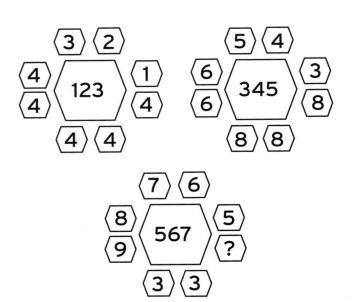

Answer see page **164**

PUZZLE 69

The squares of the times it takes planets to go round their sun are proportional to the cubes of the major axes of their orbits. With this in mind, if CD is four times AB, and a year on the planet Zero lasts for six earth years, how long is a year on the planet Hot?

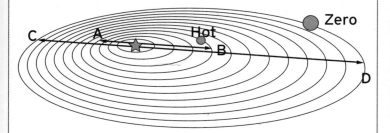

Answer see page **164**

PUZZLE 70

Fill the numbers into the blank spaces. There is only one correct way.

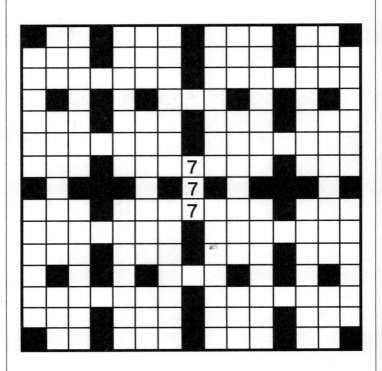

ACROSS

29	345	477	1052151
47	389	485	1285465
58	394	488	1469451
81	409	510	1779317
012	416	550	2008732
018	437	563	2457149
048	439	746	2857375
142	448	775	5125721
192	459	819	5418409
314	473	907	9588859

DOWN

138	~~777~~	1949159	6656485
198	158453	2193241	7413313
231	219952	2443740	8475941
250	420417	3854345	8614451
410	590579	4112340	8724315
473	0474542	5984178	9855707
745	1274458	6584404	9905865
750			

Answer see page **165**

PUZZLE 71

Put the stars into the boxes in such a way that each row is double the row below.

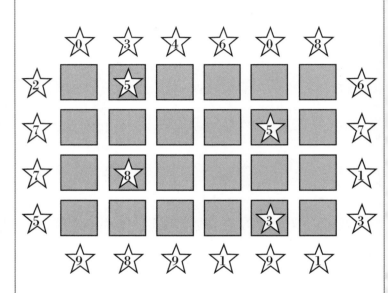

Answer see page **165**

PUZZLE 72

Which is the odd number out?

Thirty-six
Sixty-four
Seventy-two
Twenty-five
Eighty-one

Answer see page **165**

PUZZLE 73

What number should replace the question mark?

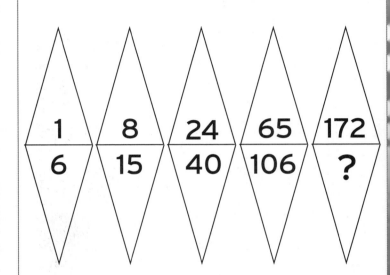

Answer see page **165**

PUZZLE 74

Insert the missing numbers to make each row, column, and long diagonal add to 189.

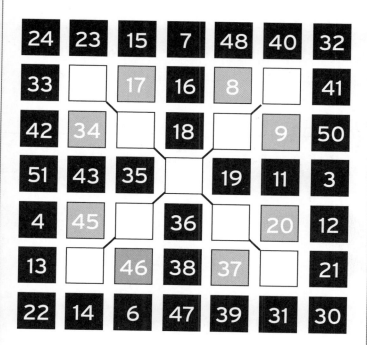

Answer see page **165**

PUZZLE 75

In 1952, New Year's Day was on a Tuesday. What day of the week was New Year's Day in 1953?

Answer see page **165**

PUZZLE 76

Find two numbers, within the digits of the number below, which give 8647492 when multiplied together.

6 5 8 8 7 2 1 4

Answer see page **165**

PUZZLE 77

What number should replace the question mark?

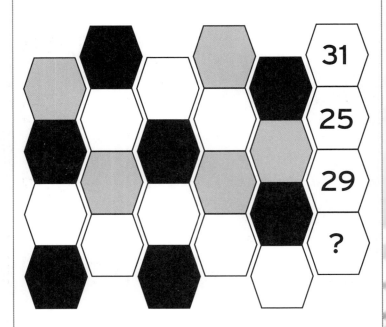

Answer see page **165**

PUZZLE 78

What is the missing number?

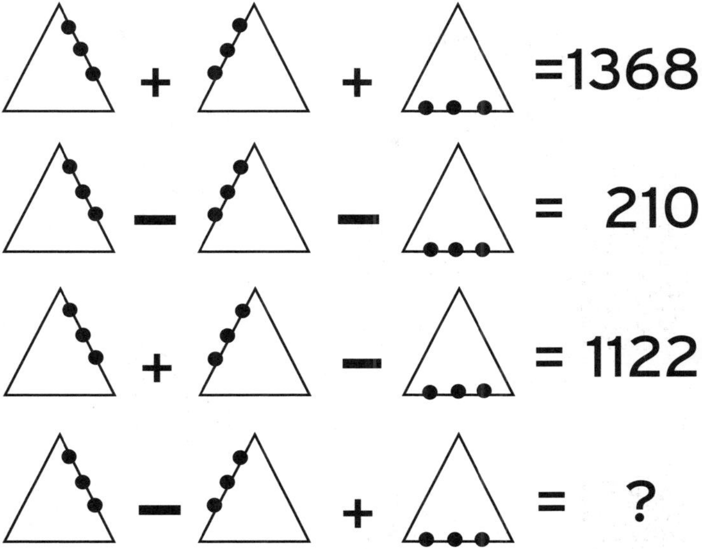

△ + △ + △ = 1368

△ − △ − △ = 210

△ + △ − △ = 1122

△ − △ + △ = ?

Answer see page **165**

PUZZLE 79

Five of these numbers interact together to give the number 1 as a solution. Which five numbers, and in which order?

+ 19	x 9	+ 29	x 7	- 999
- 94	+ 173	+ 65	- 236	x 8
+ 122	x 5	x 212	+ 577	- 567
+ 190	x 6	x 4	- 435	x 22
x 13	- 87	x 12	- 172	+ 117

Answer see page **166**

PUZZLE 80

The number below, when the digits are rearranged and multiplied by 63, produces a particularly repetitive result. What is the new number and what is the result?

17349652

Answer see page **166**

PUZZLE 81

Jon is Lorraine's brother. Diane married Jon. Diane is John's sister. Lorraine married John. Diane and Jon had seven grandchildren. Diane and Jon had three children. Lorraine and John had two children. Lorraine and John had seven grandchildren. Ricardo, one of Diane and Jon's children, and a cousin to Lorna-Jane and Frazier, did not marry, and had no offspring. Diane and Jon had two other children – Juan and Suzi. Lorraine and John had two children – Lorna-Jane and Frazier. Lorna-Jane married Juan, and had four children. Frazier married Suzi and had three children. Lorraine and John had twins. Frazier and Juan were cousins. Suzi and Lorna-Jane were cousins. Ricardo had a sister. Lorna-Jane had a brother. Frazier had a sister. Suzi had two brothers. For the above relationships to exist, how many were there, grandparents, parents, children, cousins and siblings in all?

Answer see page **166**

PUZZLE 82

Map the alphabet into two rows to work out the value of N and hence P.

$$F = 25$$

$$O = 17$$

$$L = 37$$

$$P = N + 4$$

Answer see page **166**

PUZZLE 83

Decode the logic of the puzzle to find the missing number.

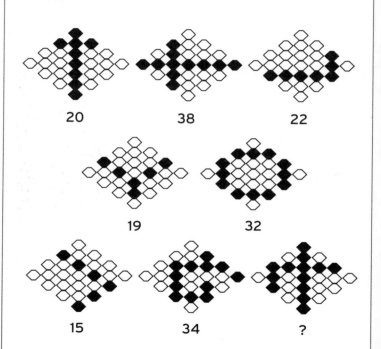

Answer see page **166**

PUZZLE 84

Insert the columns into the appropriate places to make both long diagonals add to 182.
The middle column, (D), has been done for you.

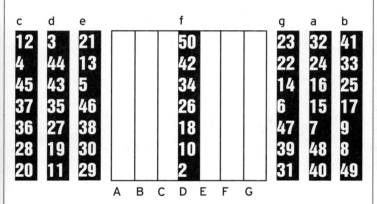

Answer see page **166**

PUZZLE 85

What is the missing number?

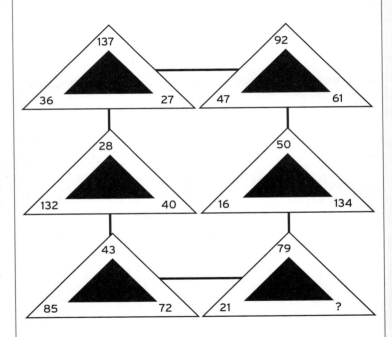

Answer see page **166**

PUZZLE 86

These systems are in balance. What is the missing number?

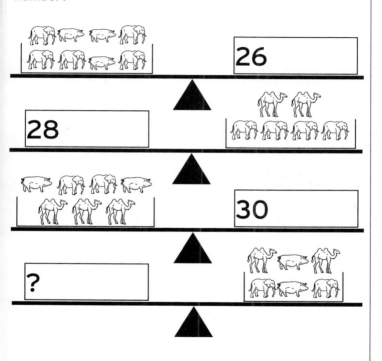

Answer see page **166**

PUZZLE 87

Find this 6-digit number.
First 3 digits − last 3 digits = 665.
Within the number there is a 3 to the left of a 1. There is a 0. There is a 7 to the right of a 9. There is a 5 to the left of a 3.

Answer see page **166**

PUZZLE 88

What is the missing number?

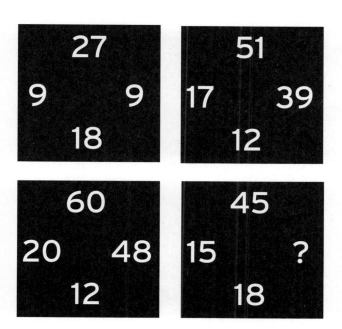

Answer see page **167**

PUZZLE 89

Put the appropriate number on the blank balloon.

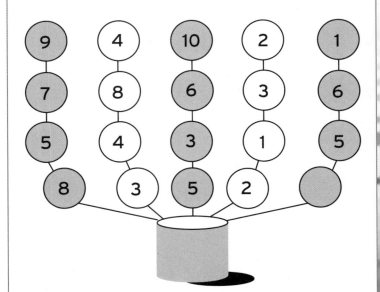

Answer see page **167**

PUZZLE 90

Fill in the blanks for Espresso.

If Abacus = 23 24 22 20

and Damascus = 52 14 22 04 22 20

then Espresso = 62 01 71 96 20 20 16

Answer see page 167

PUZZLE 91

What is the largest number you can write with three digits?

Answer see page **167**

PUZZLE 92

What is the missing number?

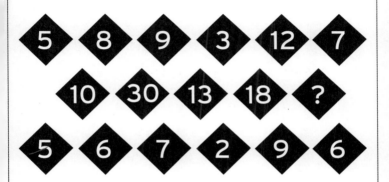

Answer see page **167**

PUZZLE 93

What is the missing number?

Answer see page **167**

PUZZLE 94

What 6-figure number, found within 3691428570, when multiplied by every number between 1 and 6, results in a number with the same digits rearranged each time?

Answer see page **168**

PUZZLE 95

Five armadillos = two pigs
One pig + one cat = one dog
One armadillo + one cat = one horse
Four pigs + two armadillos = two dogs
Four horses + three dogs = five cats + seven pigs + one armadillo

If armadillos are worth 2, what are the values of the dogs, horses, cats and pigs?

Answer see page **168**

PUZZLE 96

In the blank hexagon at the corner of each black box, write a single-digit number which, when added to the other three corner numbers, equals the total in the middle. For example, 25 could be 5 + 5 + 6 + 9. But you have to consider how the surrounding totals, 20, 19, and 21, will be affected by your choice. You must use each number – including 0 – at least once.

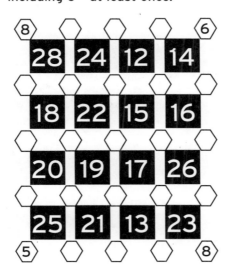

Answer see page **168**

PUZZLE 97

What is the missing number?

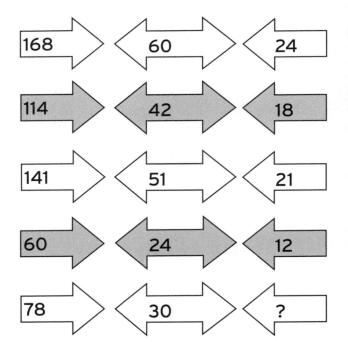

Answer see page **168**

PUZZLE 98

Put the appropriate number in the blank triangle.

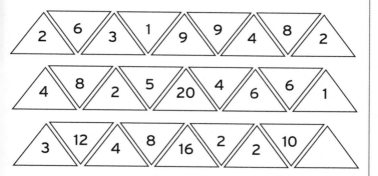

Answer see page **168**

PUZZLE 99

Find the continuous sequence of 76384 in the grid below, starting from the top row and ending on the bottom row. You may go along or down, but not up or diagonally. The numbers are not all in a straight line.

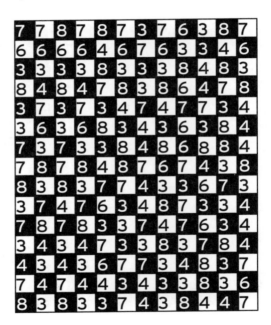

Answer see page **169**

PUZZLE 100

Hidden within the number below are two numbers which, when multiplied together, produce 1111111111111111.
What are they?

651359477124183

Answer see page **169**

PUZZLE 101

Complete the analogy.

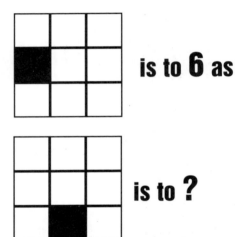

Answer see page **169**

PUZZLE 102

Fill in the blank squares.

6	0	8	3
2	9	7	7
1	0	5	3

→ | 2 | 2 | 4 |

4	5	5	5
3	2	5	2
2	7	2	0

→ | 4 | 1 | 8 |

0	3	1	6
2	5	8	9
3	9	6	6

→ | | | |

Answer see page **169**

PUZZLE 103

What is the missing number?

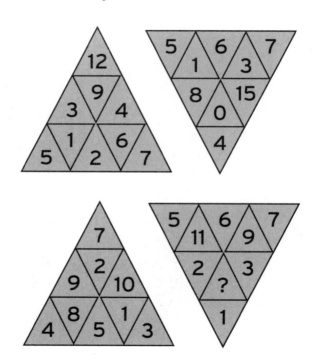

Answer see page **169**

PUZZLE 104

This system is balanced. How heavy is the black box (ignoring leverage effects)?

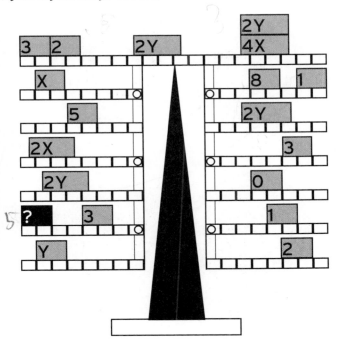

Answer see page **169**

PUZZLE 105

What is the value of the right-hand target?

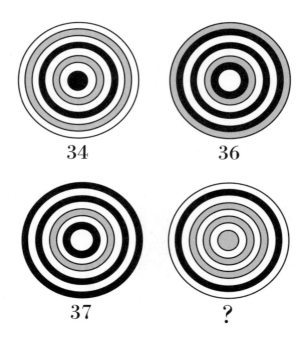

Answer see page **169**

PUZZLE 106

An architect, after drawing plans for a room, finds that if he increases the length of the room by two units, and reduces the width by one unit, while maintaining the same height of ceiling, the room will have the same volume. If the difference between the original dimensions was three units, what were the length and breadth of the room on the original drawing?

Answer see page **169**

PUZZLE 107

Use three straight lines to divide this square into five separate sections containing a total of 52 in each.

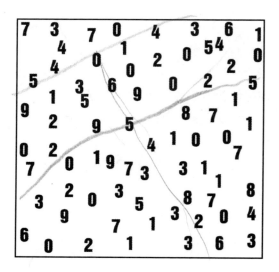

Answer see page **169**

PUZZLE 108

In four moves of two pieces, make two numbers of alternating backgrounds, such that when you subtract one from the other, the result is 671. You must not finish with any gaps between the numbers but they will occur as you work through the puzzle.

1 2 3 4 5 6 7 8

Answer see page **169**

PUZZLE 109

What is the missing number?

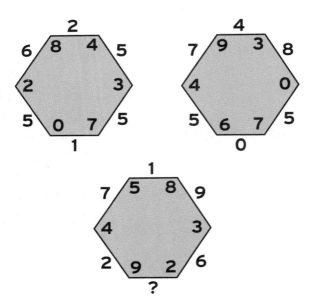

Answer see page **170**

PUZZLE 110

Which two numbers below, when you put them together and multiply the result by one of the other numbers, produces 12345678 as a result?

21 43 65 990 453 7765 8213 8890 6 5578 34 44
6012 05476 8653 9963 3257 45 75 25 2057 43567
7833 301 2134 248 54 79 92 12 38 22387 457 908
98 3245 1144 0980 356 76 91 111 88 2345 905 1121
42 5567 233 2355 8807 5467 890 20 994 1123 4356
7879 4567 67844 86743 54389 33 22 89 345665
052340 76435 345 120 243 94 123 100 53 400 335
555 613 1200 695443 2332 567 1023 845 77 325 205

Answer see page **170**

PUZZLE 111

The numbers in the three balls above each cell, when multiplied together, minus the value of the numbers in the three balls below each cell, when multiplied together, is equal to the value of the numbers inside each cell. Insert the missing numbers.

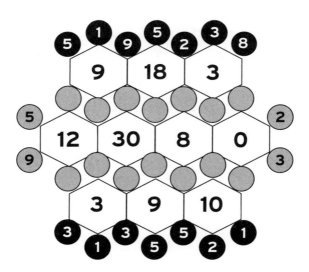

Answer see page **170**

PUZZLE 112

Insert the missing number.

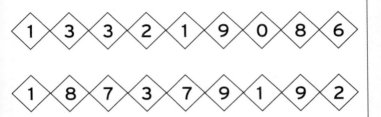

Answer see page **170**

PUZZLE 113

If each large ball weighs three units, what is the weight of each small black ball? A small white ball has a different weight from a small black ball. All small balls are solid; both the large balls are hollow.

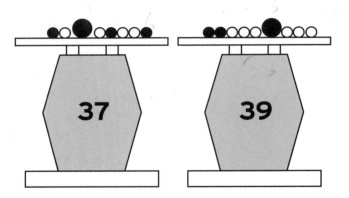

Answer see page **170**

PUZZLE 114

Each like symbol has the same value. What number should replace the question mark?

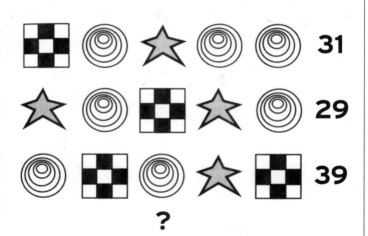

Answer see page **170**

PUZZLE 115

Insert the missing value in the blank square.

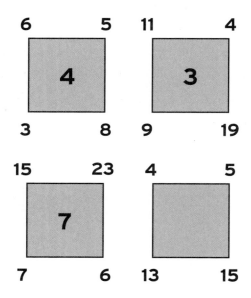

Answer see page **170**

PUZZLE 116

Which two numbers, when multiplied together, give a result that, when added to itself, produces a number that, when the digits are added together, has a solution that gives the same result as the original two numbers added together and, if doubled, produces the same result as the original two numbers multiplied together.

Answer see page **171**

PUZZLE 117

What is the missing number?

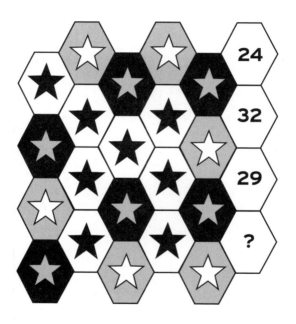

Answer see page **171**

PUZZLE 118

Put a value from below into each triangle so that the total in each square gives a value that makes each row, column, and long diagonal add to 203.

6 8 29 9 27 30 13 7 3 29 14 15 8 3 2 19 11 12 39 0 40 1 7 11 2 9 2 34 13 10 8 12 20 19 36 5 4 5 18 40

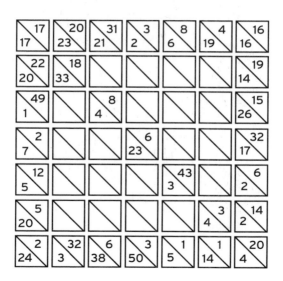

Answer see page **171**

PUZZLE 119

Insert the missing number in the blank square.

9	8	4	0
1	2	2	5

5	5	6	2
4	9	4	

8	2	7	1
3	4	2	3

Answer see page **171**

PUZZLE 120

What is the missing number?

Answer see page **171**

PUZZLE 121

Express a half, using all the digits from one to nine.

Answer see page **171**

PUZZLE 122

Decode the logic of the puzzle to find the missing number.

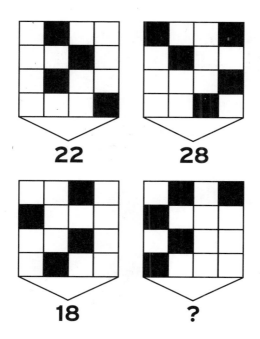

Answer see page **171**

PUZZLE 123

What are the individual values of the black, white and shaded hexagons.

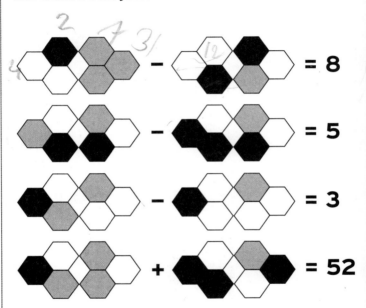

Answer see page **171**

PUZZLE 124

Somewhere within the numbers below, there is a number which, if put into the grid below, starting at the top left and working from left to right, row by row, will have the middle column as shown when the grid is completed. Put in the missing numbers.

```
1022493847460987123454668834712948876255454
4470112313501576120869252818027953987091729
3538920102603916707176981599565032900307291
8077807697853260892991202917077197830091032
5052516728962909609138507990985032910991078
27364569798236554231098467392909046229
```

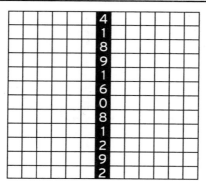

Answer see page **171**

PUZZLE 125

Find within this number, a 6-digit number which, when divided by three gives a 5-digit palindromic number. (The number is the same reading from the left and the right.)

1 3 9 1 4 0 5 9 2 1

Answer see page **172**

PUZZLE 126

Complete the analogy.

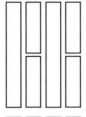 is to 24 as

 is to 28 as

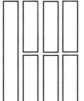 is to ?

Answer see page **172**

PUZZLE 127

What is the missing number?

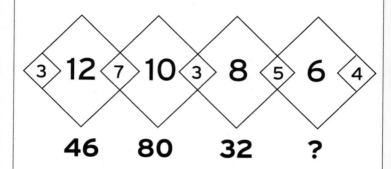

Answer see page **172**

PUZZLE 128

Which of these four sets is the odd one out?

A

1	9	0	3	3
0	3	3	1	9
	1		0	

3		9		3
9	0	3	3	1
3	3	1	9	0

B

4	2	8	1	1
8	1	1	4	2
	4		8	

1		2		1
2	8	1	1	4
1	1	4	2	8

C

3	0	9	2	2
9	2	2	3	0
	3		9	

2		0		2
0	9	2	2	3
2	2	3	0	9

D

2	4	3	5	5
5	5	2	4	3
	3		5	

4		5		2
5	2	4	3	5
3	5	5	2	4

Answer see page **172**

PUZZLE 129

Find a route from the top to the bottom of this puzzle that gives 175 as a total. Any number adjacent to a zero reduces your total to zero.

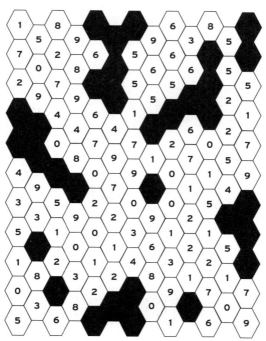

Answer see page **172**

PUZZLE 130

How many bacteria cultures should be in the Petri dish with the question mark?

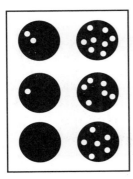
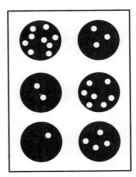

Answer see page **172**

PUZZLE 131

Insert the appropriate value in the blank triangle.

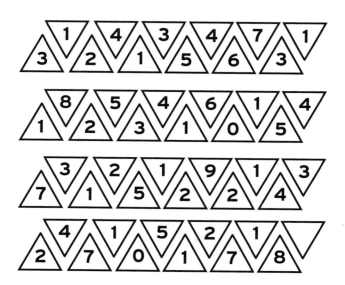

Answer see page **173**

PUZZLE 132

Which 2-digit number, the values of which when added together total 10, will always divide exactly into any 8-digit number in which the first four digits are repeated in the second half in the same order?

Answer see page **173**

PUZZLE 133

If it takes twelve men two hours to dig a hole two feet long by two feet wide by two feet deep, how long will it take six men to dig a hole twice as long, wide and deep?

Answer see page **173**

PUZZLE 134

Find two numbers so that the square of the first plus the second, added to the square of the second plus the first equals 238.

Answer see page **173**

PUZZLE 135

What is the missing number?

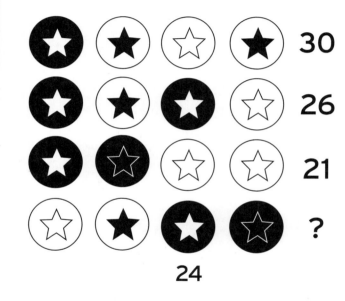

30
26
21
?
24

Answer see page **173**

PUZZLE 136

Each like symbol has the same value.
Work out the value of the missing digit.

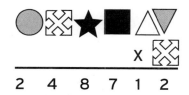

2 4 8 7 1 2

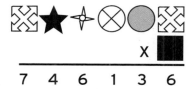

7 4 6 1 3 6

2 9 8 4 5 4 ?

Answer see page **173**

PUZZLE 137

A girl asks her mother's age and is told, "Six years ago I was nine times your age. Now I am only three times your age."

What are the present ages of the girl and her mother?

Answer see page **173**

PUZZLE 138

How many revolutions per minute does the small wheel make?

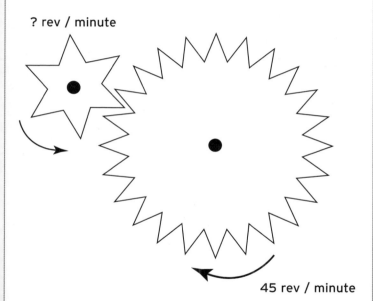

Answer see page **173**

PUZZLE 139

What is the missing point value?

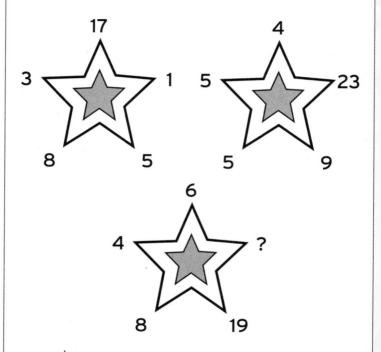

Answer see page **174**

PUZZLE 140

Use three straight lines to divide this square into five sections, each of which contains a total value of 60.

Answer see page **174**

PUZZLE 141

Use logic to find which shape has the greatest perimeter.

Answer see page **174**

PUZZLE 142

Supply both the missing numbers.

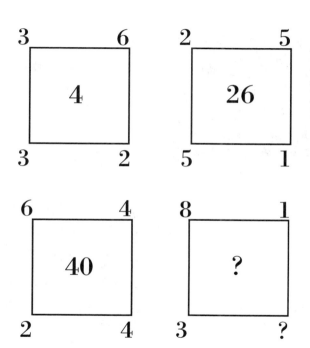

Answer see page **174**

PUZZLE 143

What number comes next in the sequence to replace the question mark?

2	7	11
20	38	69
	?	

Answer see page **174**

PUZZLE 144

What number, when you multiply it by three and multiply the result by seven and then add both of the resultant digits, and multiply the result of that by itself, when you add nineteen, gives a number that is a perfect square which can be divided exactly by only seven smaller numbers (excluding one), two of which are prime and two of which are perfect squares?

Answer see page **174**

PUZZLE 145

Make an exact quarter using all these numbers, and no other.

252525250

Answer see page **174**

PUZZLE 146

Find the missing number.

8 2 3 1 8 4
9 2 3 2 0 ?

Answer see page **175**

Answers

ANSWERS

1 A. 24. Opposite numbers are divided or added to give 24.
B. 3. Opposite numbers are multiplied or divided by 3.

2 34. Write the alphabet in a 3-row grid with the following values: A, J, S = 1; B, K, T = 2; C, L, U = 3; D, M, V = 4; E, N, W = 5; F, O, X = 6; G, P, Y = 7; H, Q, Z = 8; I, R = 9. Thus, Raphael = 9 + 1 + 7 + 8 + 1 + 5 + 3 = 34.

3 11954.
(45911 - 11954 = 33957)

4 16.

□ = 4   □ = 5
□ = 6   □ = 7

5 Carlos is oldest; Maccio is youngest. (From oldest to youngest: Carlos, Juan, Za-za, Fifi, Jorjio, Maccio.)

6 30 x 15 units (the pool's area becomes 18 x 25 units, or 450 square units).

7 25.
Circle = 4
Triangle = 8
Diamond = 5
Square = 2

8 77 square units.

9 D. The binary numbers start at the top and work left to right, line by line.

10 E, G, G. These represent the numbers 577, which are added to the sum of the previous top and middle line, to get the bottom line.

ANSWERS

11 Any number. This amazing formula will always end up with the number you first thought of, with 00 at the end.

12 400. The numbers are the squares of 14 to 21 inclusive.

13 Add 4 big balls on the right, remove 1 small from the left.

14 7 people.

15 3. Change the number of rosettes to digits and take the middle line from the top line.

16 10 people.

17 6. The right weight is nine units across to balance the left three units across. 6 x 9 (54) balances 18 x 3 (54).

18

6	2	9	3	7
3	7	6	2	9
2	9	3	7	6
7	6	2	9	3
9	3	7	6	2

19

22	21	13	5	46	38	30
31	23	15	14	6	47	39
40	32	24	16	8	7	48
49	41	33	25	17	9	1
2	43	42	34	26	18	10
11	3	44	36	35	27	19
20	12	4	45	37	29	28

20 10 m. The ratio of the flagpole to its shadow is the same as the ratio of the measuring stick to its shadow.

ANSWERS

21 D. The least number of faces touching each other gives the greatest perimeter.

22 1. A + B = KL, C + D = MN, and so on.

23 3. There are two sequences in the series: 6 × 8 = 48, and 7 × 9 = 63.

24 18.
Elephant = 2
Walrus = 3
Camel = 4
Pig = 5

25 28. Each row is a sequence of A + D = C, D + C = B and B + C = E.

26 Five men. Each man digs 1 hole in 5 hours, and thus 20 holes in 100 hours.

27 15. Add the value of the top two stars of each column to the value of the middle two stars to get the value of the bottom two stars.

28 200 Credits.
9 × 25 − (4 × 6.25) = 200.

29 21 times.

30 194. $(1 \times 5^3) + (2 \times 5^2) + (3 \times 5^1) + (4 \times 5^0)$.

31 10.
Snowflake = 5
Candle = 3
Sun = 2

32 25. Star = 9, Whorl = 5, Square = 3.

33 78. Multiply opposite numbers and add the results to get the numbers in the middle. Thus 24 + 24 + 30 = 78.

ANSWERS

34 248. Long lines = 2, short lines = 1. Add the values on the right to arrive at the answer.

35

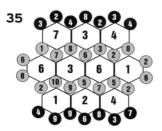

36 88. 88 + 880 + (4 x 8) = 1000.

37 27. Subtract the bottom two digits expressed as a number, from the top two digits, also expressed as a number. The difference is halved and the result is put in the middle. 78 - 24 = 54. 54 ÷ 2 = 27.

38 9 minutes and 9 seconds after 1.

39 D. The paper would reach 3,355.4432m, which is as high as a mountain.

40 B. The lighty shaded spots represent the hands of a clock. 3:00 - 9:00 = 6:00.

41 7. Take the middle number from the top left number. Multiply that by 2 to get the top right number. Add 5 to the top right number to get the bottom number.

ANSWERS

42

3		5		4		4		3		3
	90		120		64		144		54	
2		3		2		2		6		1
	48		96		16		72		36	
1		8		2		2		3		2
	160		80		20		150		30	
4		5		1		5		5		1
	180		10		40		100		15	
9		1		2		4		1		3
	27		8		32		12		81	
3		1		4		1		3		9
	24		28		84		45		135	
8		1		7		3		5		1
	144		42		63		225		25	
3		6		1		3		5		1

43 42. The bottom number goes next to the top one to make a two-digit number; the left and right do the same. Then subtract the second number from the first. 96 − 54 = 42.

44 32.
Diamond = 7
Circle = 4
Hexagon = 13
Square = 8

45 38 seconds after 8.43.

46 7. There are 7 areas of intersection at this position.

47 12, 19, 26, 3, 10. The bottom line of a Magic Square, in which all rows, columns, and long diagonals equal 70.

48 0. In all the shapes the top two numbers are multiplied, then halved. NB 3 x 0 = 0.

49 4. Start from the top left of the spiral and work in, successively subtracting and adding: 9 − 7 = 2, 2 + 5 = 7, etc.

50 103.5.

ANSWERS

51 19. They denote the alphanumeric positions of numbers from 1 to 6. The first letter of six is "s", the 19th letter of the alphabet.

52 There are 5 cards missing, leaving 47 in the deck.

53 22. Hexagon = 8; Cross = 3; Diamond = 2

54 2. C = A − B, with the result reversed. 496324 − 235768 = 260556.

55 19. The top pair of numbers are multiplied together and added to the result of multiplying the bottom pair of numbers together. (2 x 8) + (3 x 1).

56 A. 54, B. 42. Opposite numbers are multiplied, divided, or added to get the numbers in the middle.

57 1. The number is an anagram of Mensa, with numbers substituted for the letters.

58

```
    912           921
   x 36          x 36
   ----          ----
   5472          5526
   2736          2763
   ----          ----
  32832         33156
```

59 3 units. The difference of 24 divided by 8.

60 96. $4^2 = 16$; 16 x 6 = 96.

ANSWERS

61

6	8	0	9	4	1	**6**	4	1	6	2	2	2
3	4	5	6	3	4	**1**	2	1	9	1	8	3
6	9	1	6	1	4	**4**	4	3	2	7	0	8
9	2	2	8	4	6	**1**	5	2	9	5	5	0
0	1	6	2	1	9	**3**	2	0	0	2	5	
2	8	1	3	1	2	**1**	5	8	5	8	7	1
9	3	9	4	5	0	**4**	6	3	9	5	1	2
3	1	6	1	7	6	**2**	1	1	3	2	6	7
7	9	2	2	8	9	**6**	5	6	1	2	3	1
0	2	2	3	8	4	**0**	4	6	1	2	8	9
8	5	4	0	4	3	**2**	6	1	6	1	4	2
5	2	6	1	6	0	**9**	3	4	1	7	2	8

62 24 ways. There are six alternatives with each suit at the left.

63 B. Each nodule is given a value, depending on its position in the grid. The values are added together.

64 15:03 (or 03.15 (pm) if the watch has the capacity to switch to 12-hour mode).

65 7162 and 3581.

66 2. The weight is positioned 8 units along, so it needs a weight of 2 units (8 x 2 = 16) to keep the system in balance.

67
A = 5. (a + b) − (d + e) = c
B = 0. (d + e) − (a + b) = c
C = 3. a + b + c − e = d
D = 2. c + d + e − a = b

68 2. The top four numbers, plus the number in the middle, equals the bottom four numbers. Hence 8765 + 567 = 9332.

69 9 earth months. Zero has an orbit that takes $\sqrt{4^3}$ (8) times as long as Hot.

ANSWERS

70

71 ☆5 ☆6 ☆3 ☆ ☆2
☆ ☆8 ☆1 ☆ ☆6
☆8 ☆9 ☆0 ☆ ☆8
☆9 ☆4 ☆ ☆5 ☆ ☆9

72 72. It is the only non-square number.

73 279. The numbers are added together and the sum + 1 is put in the next triangle. 106 + 172 = 278 + 1 + 279

74

75 Thursday. 1952 was a leap year with 366 days. 366 ÷ 7 (days in a week) = 52 remainder 2. Tuesday + 2 days = Thursday.

76 1258 x 6874.

77 19.
Shaded = 9
Black = 5
White = 3

78 456. The first symbols are worth 789; the middle symbols are worth 456; the right-hand symbols are worth 123.

ANSWERS

79 + 29, x 7, - 94, x 4 and - 435. The sum is: 29 x 7 (203) - 94 (109) x 4 (436) - 435 = 1.

80 12345679 (x 63) = 777777777.

81 16 people.

82 P = 19. Map the alphabet into 2 rows of 13 each. Then add the numerical values of each row to get the value of the letters. A (1) + N (14) = 15. P comes two letters after N in the alphabet, so add two to the top and two to the bottom (16 + 3 = 19).

83 35. White hexagons have no value. Black hexagons are worth 1 in the top row, 2 in the second row, 3 in the third row, 4 in the fourth row, then 3 in the fifth row, 2 in the sixth row and 1 in the seventh row.

84

e	c	d	f	b	a	g
21	12	3	50	41	32	23
13	4	44	42	33	24	22
5	45	43	34	25	16	14
46	37	35	26	17	15	6
38	36	27	18	9	7	47
30	28	19	10	8	48	39
29	20	11	2	49	40	31
A	B	C	D	E	F	G

85 100. The numbers inside each triangle total 200.

86 24.
Elephant = 4
Pig = 2
Camel = 6

87 975310.

ANSWERS

88 27. The left number is one-third of the top and the right is subtracted from the top number to give the bottom.

89 4. In each row, the numbers in the two left balloons equal the numbers in the three right balloons.

90

(62) (01) (71) (96) (20) (20) (16)

Move one place to the right in the alphabet. A = 2, B = 3. The numbers to make espresso are E = 6, S = 20, P = 17, R = 19 and O = 16.

91 $9^{(9^9)}$ (nine [to the power of nine, to the power of nine]). Solve the top power first, giving nine to the power of 387420489. The result is a number so large that it has never been calculated.

92 30. Multiply adjacent numbers in the top line together and adjacent numbers in the bottom line together. Then subtract the lower from the higher and put answer in the middle. This is done continuously. (12 x 7) [84] − (9 x 6) [54] = 30.

93 252.
Black triangle = 6
White triangle = 3
12 x 21 = 252.

ANSWERS

94 142857.
The numbers are:
(x 1) 142857
(x 2) 285714
(x 3) 428571
(x 4) 571428
(x 5) 714285
(x 6) 857142

95 Dog = 12
Horse = 9
Cat = 7
Pig = 5

97 14. Divide the left number by 3 and add 4 to give the middle number. Repeat the sums with the middle number to get the right number. 78 ÷ 3 = 26; 26 + 4 = 30; 30 ÷ 3 = 10; 10 + 4 = 14.

98 5. Each line contains three separate multiplication sums with the answer in between the multipliers. 10 ÷ 2 = 5.

96

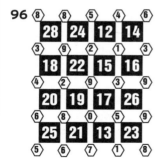

99

ANSWERS

100 17 x 65359477124183 = 1111111111111111.

101 8. The squares are numbered from 1 to 9, starting on the top left, from left to right, right to left, left to right.

102 425. Reverse the top line, subtract the second line from that, and subtract the result from the bottom line to get the three figure sum for the blanks. 6130 - 2589 = 3541; 3966 - 3541 = 425.

103 5. The numbers in all the triangles add up to 49.

104 2 units. The difference of 18 divided by 9. The units to the right come to 104, to the left they are 86. 104 - 86 = 18. The blank box is 9 units across so 2 x 9 = 18.

105 31.
White ring = 4
Black ring = 6
Shaded ring = 3

106 It was 8 x 5 units. This becomes 10 x 4 units, retaining the area of 40 square units.

107

7	3	7	0	4	3	6	1
	4		1	0	2	0	5 4	0
5	4		6	9	0	2	2	5
9	1 3 5	5		8	7	1	
	2		2		4	1 0	0	7
0	7	0	1 9	7 3	3 1	1	
	3	2	0	3	5	8 7		8
6		9		7	1	3 2 0	4
	0	2	1		3	6	3

108 The sum is 7153 - 6482 = 671. Move the pieces as follows (overleaf):

ANSWERS

1	2	3	4	5	6	7	8
6	7	1	2	3	4	5		8
6	7	1	2			5	3	4	8
6			2	7	1	5	3	4	8
6	4	8	2	7	1	5	3

109 6. In each case, the sum of the numbers outside a hexagon equals the sum of the numbers inside it.

110 2057, joined by 613, x 6. The sum is 2057613 x 6 = 12345678.

111
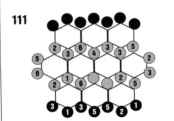

112 5. The first three digits expressed as a whole number, subtracted from the next three digits, expressed as a whole number, equals the last three digits. 623 - 188 = 435.

113 Small black balls weigh 6 units. White balls weigh 4 units.

114 21. Whorl = 5
Checkered box = 13
Star = 3

115 3. Add the left top and bottom numbers together, add the right top and bottom numbers together. Subtract the smaller from the larger to get the middle number. (15 + 5) - (13 + 4) = 3.

ANSWERS

116 6 and 3. 6 x 3 = 18; 1 + 8 = 9 (and 6 + 3 = 9); 9 + 9 = 18.

117 27.
Shaded hexagon/ white star = 3
Black hexagon/ shaded star = 5
White hexagon/ black star = 8

118

17/17	20/23	31/21	3/2	8/6	4/19	16/16
22/20	18/33	11/0	10/3	9/13	29/2	19/14
49/1	5/5	8/4	2/19	27/3	30/9	15/26
2/7	14/4	12/8	6/23	18/20	29/11	32/17
12/5	7/12	13/15	36/1	43/3	40/8	6/2
5/20	19/8	2/34	39/6	7/40	43/4	14/2
2/24	32/3	3/38	3/50	1/5	1/14	20/3

119 7. In each box, top left x bottom right = bottom left and top right. The products are a two-digit number reading up.

120 3. Reverse the second line and subtract it from the top line to get the bottom line. 43390 - 25587 = 17803.

121 $\dfrac{6729}{13458}$

122 20. In each shape, the values are of the black squares. In column 1, they are worth 2; in column 2, they are worth 4; in column 3, they are worth 6 and in column 4, they are worth 8. The values are added together and the total goes at the bottom.

123 Black hexagon = 2
White hexagon = 4
Shaded hexagon = 7

ANSWERS

124

```
8 8 7 6 2 5 5 4 5 4 4 7 0 0 1
1 2 3 1 3 5 0 1 5 7 6 1 2 0 8
6 9 2 5 2 8 1 8 0 2 7 9 5 3 9
8 7 0 9 1 7 2 9 3 5 3 8 9 2 0
1 0 2 6 0 3 9 1 6 7 0 7 1 7 6
9 8 1 5 9 9 5 6 5 0 3 2 9 0 0
3 0 7 2 9 1 8 0 7 7 8 0 7 6 9
7 8 5 3 2 6 0 8 9 2 9 9 1 2 0
2 9 1 7 0 7 7 1 9 7 8 3 0 0 9
1 0 3 2 5 0 5 2 5 1 6 7 2 8 9
6 2 9 0 9 6 0 9 1 3 8 5 0 7 9
9 0 9 8 5 0 3 2 9 1 0 9 9 1 0
```

125 140592. 140592 ÷ 3 = 46864.

126 20.
Long bar = 8
Short bar = 2

127 39. Each diamond contains three numbers. To get the bottom number, multiply the left by the middle, and add the product to the sum of the right and the left.
(5 x 6) + 5 + 4 = 39.

128 D. The two sections of each shape fit together to form a magic square. Each row of the other three add to 16, but each row of D adds to 19.

129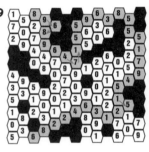

130 8. In each box the top two dishes expressed as numbers are multiplied to give the middle two dishes. The middle two dishes are then multiplied in the same way to give the bottom two.
6 x 4 = 24; 2 x 4 = 8.

ANSWERS

131 2. Each row adds to 40.

132 73.

133 32 hours. The hole will have eight times the volume. It would take 12 men eight times as long to dig it, and 16 times as long for 6 men.

134 7 and 13. 7^2 (49) + 13 = 62. 7 + 13^2 (169) = 176. 62 + 176 = 238.

135 23. The shapes have the following values:

7 **9** **5** **2**

136 4. The sums are
124356 x 2 = 248712
248712 x 3 = 746136
746136 x 4 = 2984544
The shapes have these values:

1 **2** **3** **4**

△ ▽ ○ ✦
5 **6** **7** **8**

137 The girl is now 8, and her mother is 24.

138 180 revolutions. (45 revolutions x 24 teeth of big wheel [1080 movements]) ÷ 6 (teeth of small wheel) = 180.

ANSWERS

139 1. The sum of the four smallest values equals the largest value. The largest value rotates by one turn clockwise each star.

140

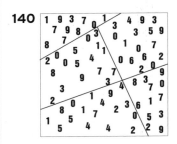

141 A. The thinnest shape to cover an area always has the greatest perimeter.

142

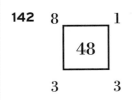

In each box, multiply the two bottom numbers and square the product to get the two top numbers. Read the top and bottom numbers as 2-digit figures and subtract the smaller from the larger. √81 = 9; 9 ÷ 3 = 3; 81 - 33 = 48.

143 127. Three adjoining numbers are added together in a continuous string. 20 + 38 + 69 = 127.

144 3. 3 x 3 [9] x 7 [63]; 6 + 3 [9] x 9 [81]; 81 + 19 = 100. 100 is 10^2 and it is divisible by 2 (prime number), 4 (also 2^2), 5 (prime number), 10, 20, 25 (also 5^2) and 50.

145 $\dfrac{5555}{22220}$